# The Gestalt Therapy Primer

by
DANIEL ROSENBLATT

*Illustrations by Charles Morris Mount*

PERENNIAL LIBRARY
Harper & Row, Publishers
New York, Evanston, San Francisco, London

*For Naomi*

THE GESTALT THERAPY PRIMER

Text copyright © 1975 by Daniel Rosenblatt. Illustrations copyright © 1975 by Charles Morris Mount. All rights reserved. Printed in the United States of America. No part of this book may be used or reproduced in any manner without written permission except in the case of brief quotations embodied in critical articles and reviews. For information address Harper & Row, Publishers, Inc., 10 East 53d Street, New York, N.Y. 10022. Published simultaneously in Canada by Fitzhenry & Whiteside Limited, Toronto.

*Designed by Eve Kirch Callahan*

First PERENNIAL LIBRARY edition published 1976

LIBRARY OF CONGRESS CATALOG CARD NUMBER: 75-18906

STANDARD BOOK NUMBER: 06-080353-3

76 77 78 79 80 10 9 8 7 6 5 4 3 2 1

# Contents

*Introduction*     1

1. Here and Now     13
2. How It All Comes Together     37
3. Learning to Be Myself, Not Who I'm Supposed to Be     53
4. What I Can Learn About Myself When I Get Angry and Blame Others     95
5. How I Can Avoid Turning Against Myself and My Body     137
6. What Happens Now?     181

*Appendix: A Brief Note About the History of Gestalt Therapy*     188

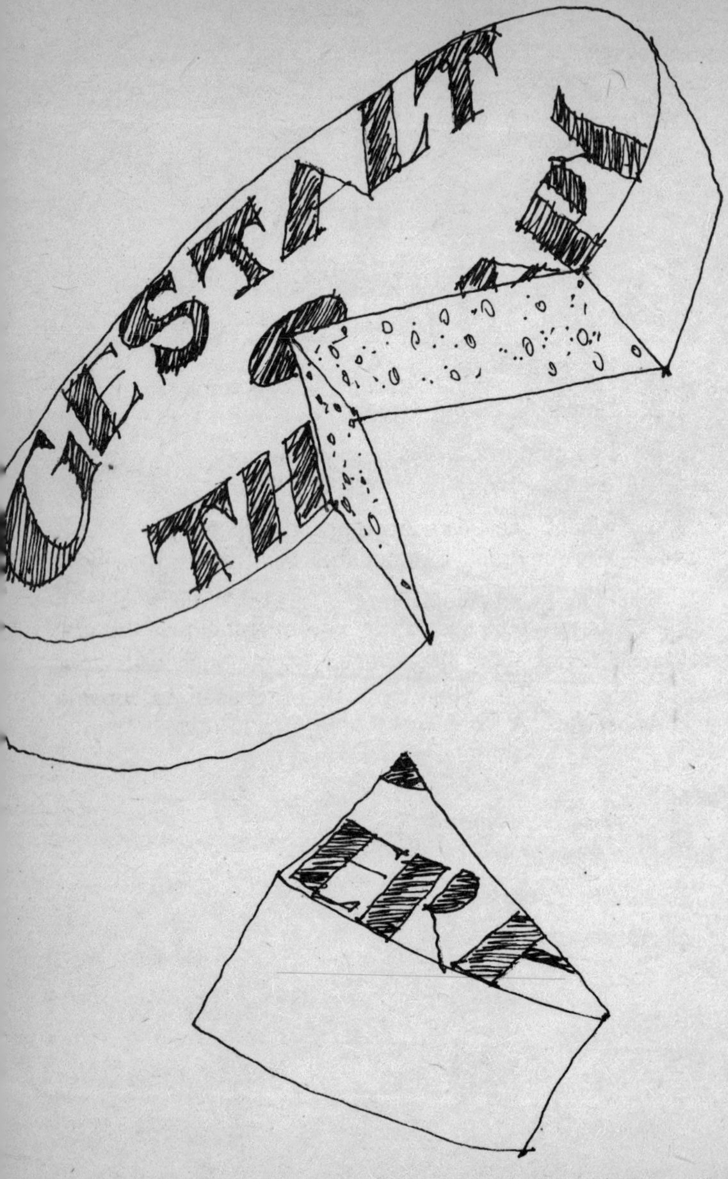
GESTALT
TH

# Introduction

This book is intended as a first taste, a small bite, a slice of what gestalt therapy is like.

In a real session with a gestalt therapist, what happens is cooked up for just you.

Here, since I don't know you, I will invent what might happen between us, so that you can gain some idea of the kind of thinking and the kind of techniques that are used to help you to grow, to get in touch with your feelings, to explore hidden parts of yourself, to become whole.

Here are several ways a gestalt therapist behaves:

He is likely to:

> be active
> talk
> suggest experiments
> comment on what is happening right now in the session
> make few interpretations
> express his own feelings to you
> make therapy a partnership
> ask you to take risks
> be especially interested in *how* not *why*
> be more interested in your present, rather than your past
> pay attention to what you do with your body:
>
>> how do you breathe?
>> where do you feel tense?
>> how do you walk?
>> how do you hold yourself?

direct your attention away from your words to your body; to ask things like:

>how does your mouth feel now?
>can you feel your forehead?
>what if your foot could talk? what might it say now?
>can you tell me how you are breathing?

ask you to take responsibility for what happens to you:

>what is good about being sick?
>what is good about being angry with your father?
>what is good about getting fired?
>what is good about oversleeping?

A gestalt therapist is likely to:

ask you to consider that in large measure, what happens to you is because you *make* it happen, even if at this moment you are not aware of *how* you make it happen

and even if you are not *aware* of what you get out of this happening, since it may look as if you get nothing out of being sick, angry, oversleeping, and so forth.

If you examine this "nothing" more closely, you will likely discover that there is a lot to be gotten out of what looks so negative.

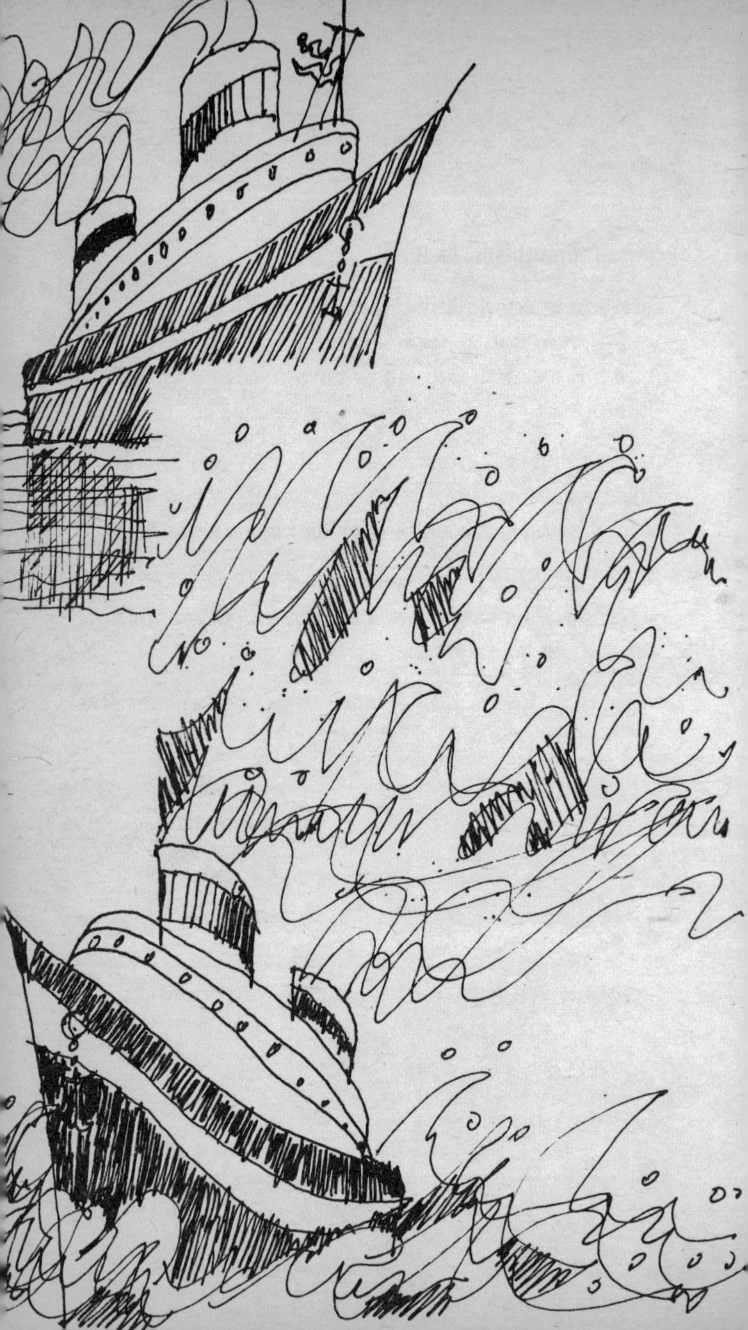

A gestalt therapist is likely to:

treat dreams in a different way.

He may be more interested in what is happening between the two of you than in your dream.

He may suggest that you dreamed up your dream to lead you or him astray.

He may ask you to become each element in the dream:

(I dreamed I was lost on a huge ocean liner with large waves slapping the ship.)

He may ask you to become a large ocean liner and let it speak.

He may ask you to become the large waves slapping the ship, and let them speak.

He may ask you to invent a dialogue between the ship and the waves.

He may let you discover the meaning of your dream, and not offer his own interpretation.

Now, if this first taste of gestalt therapy pleases you, then you might want to go further.

One way is to read *Gestalt Therapy* written by Fritz Perls, Paul Goodman, and Ralph Hefferline in 1951. The other is to find a good gestalt therapist and explore more ways to grow.

You decide for yourself. This is part of gestalt therapy, too, the belief that, ultimately, you know what is best for you. As a gestalt therapist, I want to enable you to come to this knowledge yourself and to use it.

Here are two suggestions (you can ignore them) about this book.

This book is deliberately small and short. That is because you are asked to do the experiments by yourself.

If you put down the book and do the experiments, much more will happen.

Find your own way to use this book. You do not need to do each experiment in order, you do not need to read the book in one sitting, you do not need to like each experiment.

If some experiments are too hard, skip them and come back to them at a later time.

# HERE and NOW

# 1

## Here and Now

What are you FEELING right NOW?

What are you THINKING right N−O−W?

What are you DOING right N−O−W?

What is happening *right now, here* in this situation, between the two of us?

Try an experiment. Put down this book. Then begin to pay attention to what happens to you. DO IT NOW!

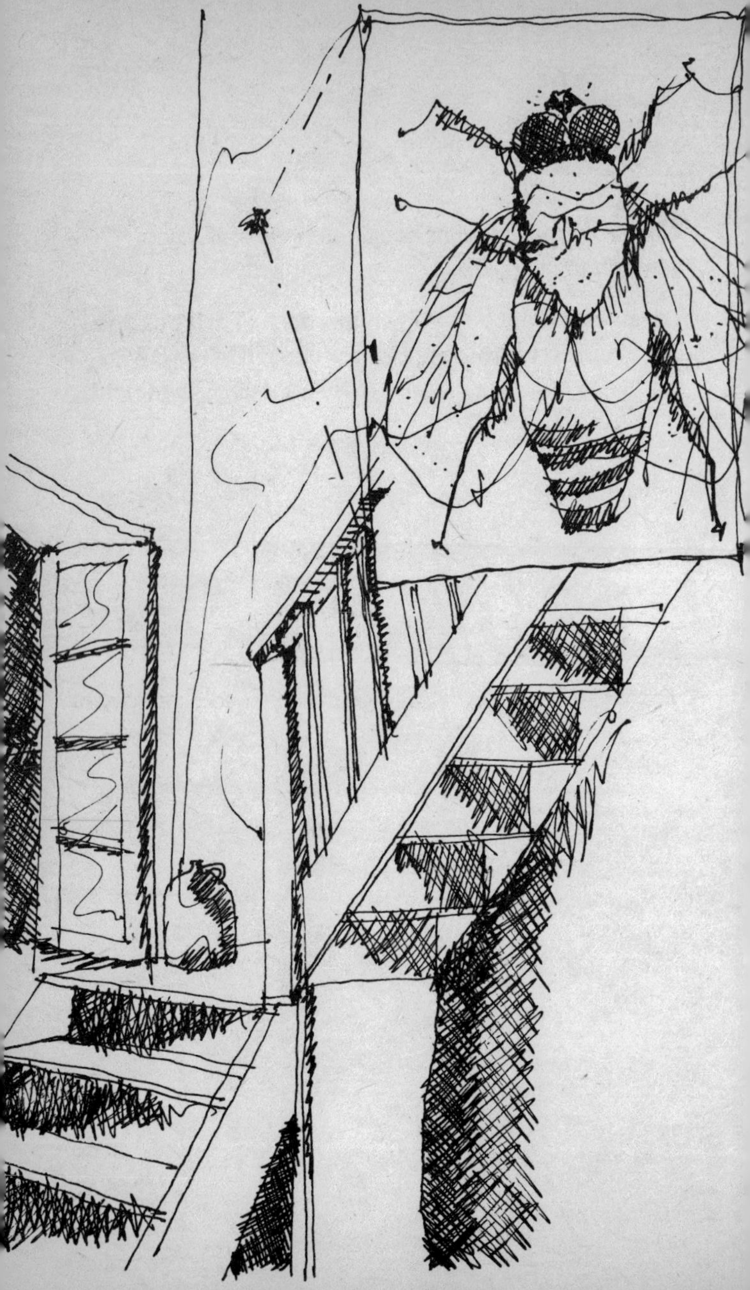

What happened?

  Here's a whole host of possibilities:

    Nothing happened. I just felt stupid!

    I felt this was a dumb book and a silly assignment.

    I thought about going to a movie.

    I had a fantasy about meeting a wonderful person.

    I noticed a fly that was crawling on the wall.

    I felt all funny inside.

    I began to get a headache.

    I began to think about what happened yesterday at work.

    I began to think about all the things I had to do today.

    . . . . . . . . .

let it be

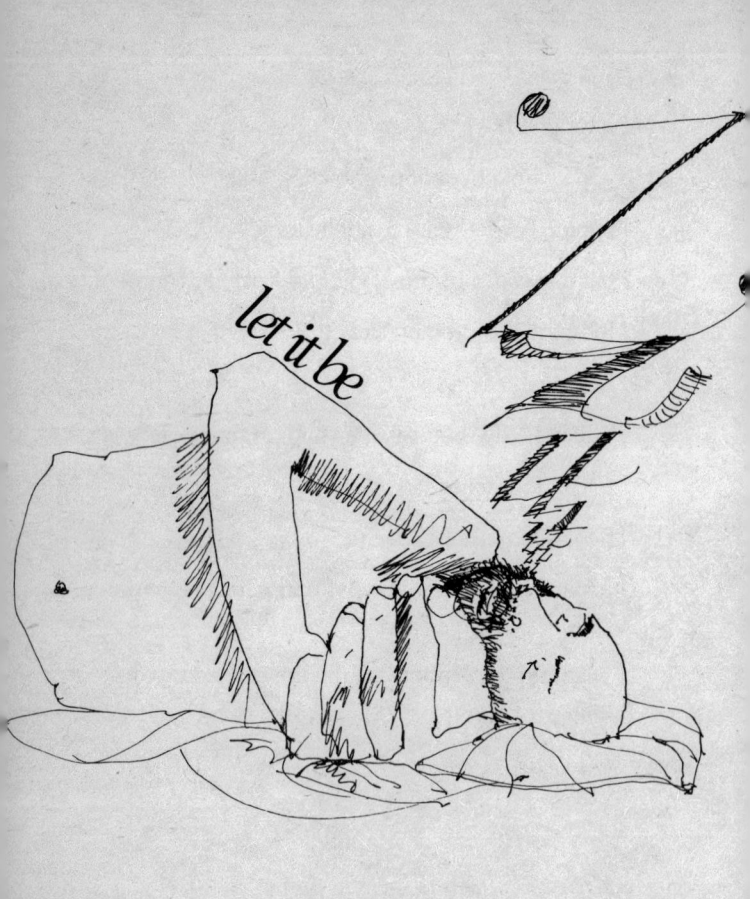

Now try something else.
Something very difficult.

Try to ACCEPT whatever it was that happened to you, to just let it be.

Try *not* to criticize, blame, or be guilty about what happened. Try to let what happened be just that—*whatever it was*.

Try NOT to JUDGE whatever it was that happened to you. See if you can withhold taking a stand about it. See if you can look at what happened without deciding: Is it *good* or *bad?*

Try to say, "That was interesting," even if you don't *feel* it or *believe* it.

criticism

Can you recognize that you might be *judging* (blaming, criticizing) yourself harshly?

or *judging* (blaming, criticizing) this book in the same harsh way?

or judging others the same harsh way?

Does your criticism interfere with your getting involved with yourself and others? your work?

Let us experiment with your criticism for the moment. Just lay it aside.

Let me give you permission not to criticize either yourself or me.

Can you accept this permission?

How long can you keep it before you start criticizing again?

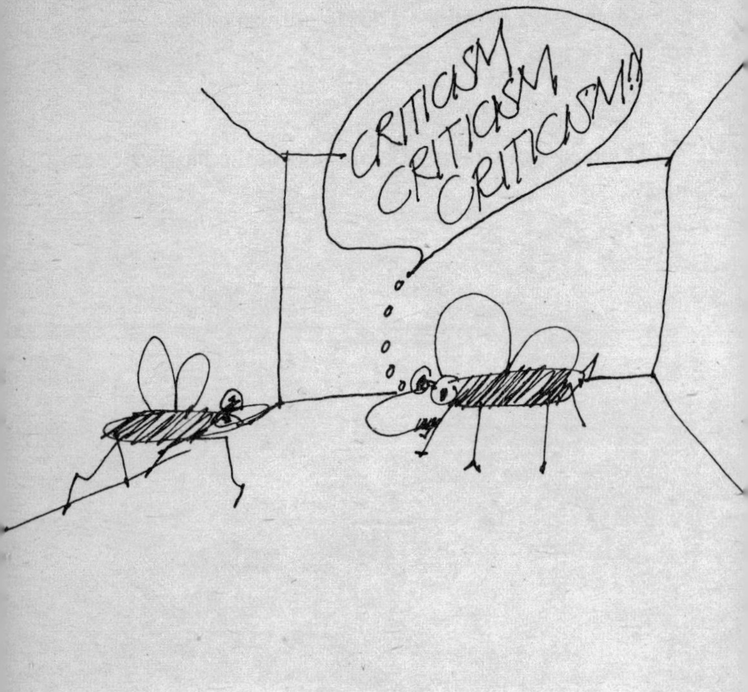
CRITICISM
CRITICISM
CRITICISM!

Do you find you *need* to criticize?

Many people do. This often becomes their style of living.

Right now it is enough to know that your criticism may interfere with your being aware of what else happens to you.

# RIGHT NOW

Now try again to find out what you are *aware* of right *now;* what you are

> feeling?
> thinking?
> doing?

Work at it for a little bit.
Concentrate.
Pay attention to yourself.
Be aware of whatever is happening.

What happened this time?

　　. . . . .
　　　. . . . .
　　　　. . . . .

It doesn't make any difference *what* happened. What is important is that some thing (something) happened. Can you treasure whatever happened?

How quickly we ignore what happens, how we overlook it, forget it, trivialize it, criticize it!

And yet, that is all we can really be sure of, the right now.

What is passed, past, is already changed, perhaps forgotten, certainly altered by memory. And what is to come, the future, is uncertain, unreal, only a possibility.

Yet, too often, *right now* is thrown away.

> Isn't it interesting that, when I have a chance to pay attention to what is happening right now, I go blank; I find nothing is happening.
>
> I want to know more about my blank-ness; I want to know more about how I make nothing happen.

I am now beginning to find out something about myself: I blank out my *now*.

And I am beginning to set up a situation where I can learn more about it.

In other words, although it is only in a small way, I have begun to get to know myself and to open the possibility of growth.

You have just had a small taste of what can happen for you.

But, please, do not mistake the taste for a whole meal.

In the pages that follow I want you to taste other aspects of your self, of your life. But, please, remember this is just a taste, just a touch, just a beginning.

If your appetite is whetted, if you feel hungry for more, then you should find a qualified gestalt therapist to continue working with.

To try to continue to grow and develop through just the experiments and suggestions in this book is to limit drastically what can happen for you.

# 2

## How It All Comes Together

shape
shape
shape
shape
whole
whole
whole
whole
form
form
form
form
configuration
configuration
configuration
configuration
theme
theme
theme
theme
structure
structure
structure
structure
structure
structure
structure
structure
theme
theme
theme
theme
configuration
configuration
configuration
configuration
form
form
form
form
whole
whole
whole
whole
shape
shape
shape
shape

*Gestalt* is a German word for which there is no single English equivalent.

In German, the word *Gestalt* is similar to, but not the same as, a whole range of English words:

> form
> shape
> whole
> configuration
> structure
> theme

Gestalt means the way in which the pieces are arranged or organized into a whole.

Thus a person is a person because he has a head, a torso, and limbs organized in a particular way.

If he had arms where his head was supposed to be, we would not call him a person, but a freak.

g
ge
ges
t
ta
tal
talt
**gestalt**

So our gestalt for physical person includes not only parts of the body, but the demand that they be arranged in a certain way that has *meaning* for us.

Similarly, a tree possesses

> roots
> trunk
> branches
> leaves

When these are all together properly, a new whole is created from these parts: a tree.

It is one of the properties of gestalts that the whole is more than the sum of its parts. The tree is more than just its roots, trunk, branches, leaves. A tree involves the form that all of these parts make; how they are put together.

To be aware of a gestalt, we have to arrange the parts so that a meaningful whole is created.

So in forming gestalts we are continually creating our experiences and giving meaning to what is happening to us. It is this process that is looked at so closely in therapy.

For instance:

> If I want to experience my job, to find out its meaning, to create a new whole, I might ask what it is, and what it is not; what I like about it and what I do not like; what jobs I gave up for this job and what jobs might still be open to me.

Let's see what happens.

## MY JOB AS A GESTALT THERAPIST

*What it is*

> a hard job
> an exciting job
> an important job
>
> it takes place indoors
> it takes concentration
> it pays well
>
> I work with people
> I work with people who have hard times in their lives
> I work with people who want to find out more about themselves

*What it is not*

> I do not use my body as much as my mind
> I sit rather than run or jump
> I do not get happy patients
> I don't often get bored

## MY JOB (continued)

*What I like about it*

> I feel useful to others
> I believe I do it well
> I am proud of my ability
> I think it is important work

*What I don't like about it*

> I have to talk a lot
> I have to sit indoors a lot
> I am close to so much pain, misery, suffering
> I ask people to give up precious illusions
> When people feel more whole, enjoy life, then
>     it is time to stop my work with them

*What jobs I have given up that I like*

> farming
> breeding dogs and cats
> running a camp
> working with infants

*What other jobs I can do at the same time*

> write
> lecture
> teach
> supervise

After I have looked at all of the parts of my job, a new figure begins to emerge from all of the parts. I discover new meanings in what I am doing and what I might want to do. Even if I decide to do nothing new, to make no changes, I still have a better feeling of what I am doing, about who I am.

And if I decide I want to make any changes, then I have a better grounding for what I want to do, of where I want to go.

In other words,

I now have a strong, sharp gestalt about my job.

Usually what goes along with a clear gestalt is a sense of excitement, a feeling of aliveness, a potentiality for growth.

This is how I *feel* right *now* about my job.

Now suppose I found out that, on the whole, I didn't like my job, and I have kept this hidden. Now I know it.

I would still have the sense of excitement, because I would be clear that I was unhappy with my work. I would look at the other kinds of work that are possible.

I would begin with a sense of possibility, to consider in what direction I want to change my job.

I might not have an answer right away, but I would have the support I need to take the next step, to explore where I might want to go.

So what I explored at one moment becomes the background of a new gestalt: What do I like about my job has developed into what job will I look for now?

GESTALT GESTALT

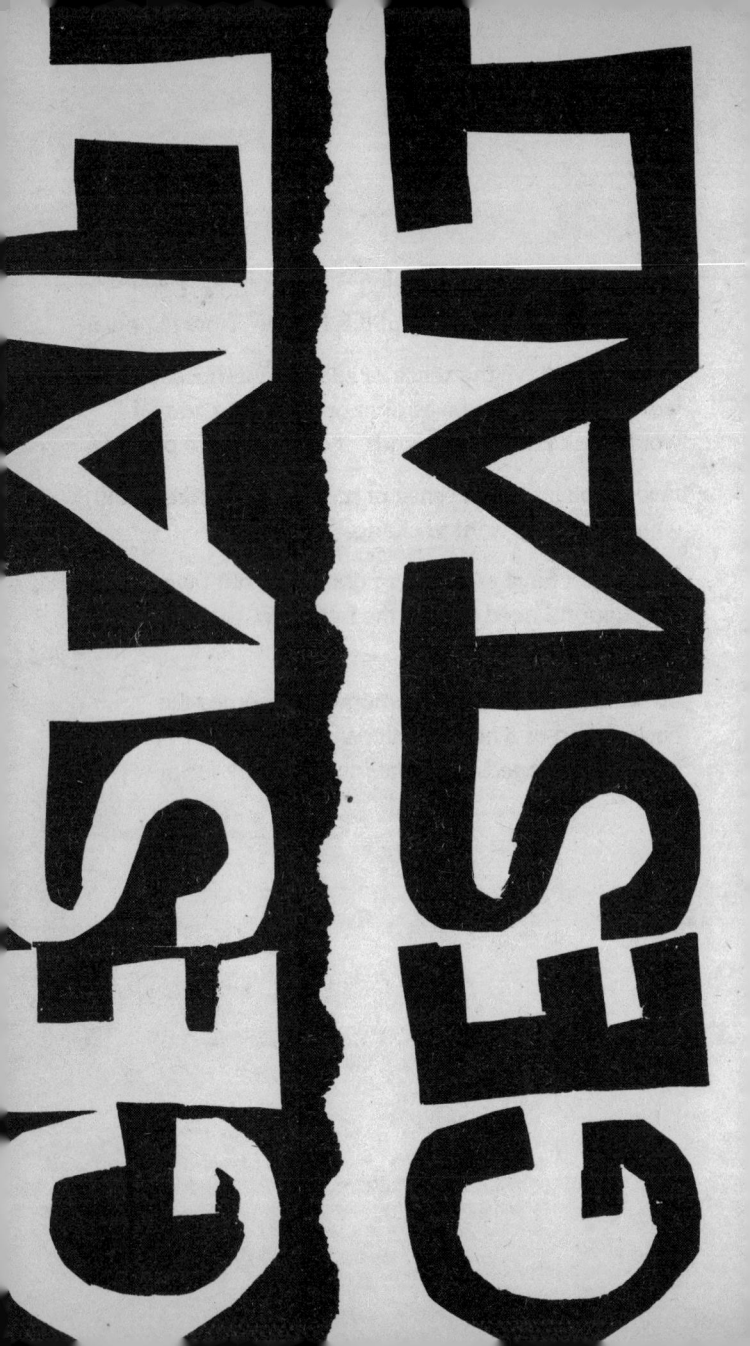

Now you can see why the process of forming gestalts is so important in growth and development.

One gestalt leads to another.

If I can be free enough to allow new gestalts to form, then my life will constantly change as my experience changes, as new meanings emerge, as new developments occur.

If I hold on to gestalts, keep them fixed or get stuck with them, then my life becomes boring, repetitive, unreal, safe.

If I leap into new gestalts too quickly, before they have clearly emerged, then my life becomes unsure, inauthentic, reckless, tense.

So how I deal with gestalts is a very important part of determining what my life will be like.

# 3

## Learning to Be Myself, Not Who I'm Supposed to Be

Since forming new gestalts is so exciting and leads to growth, why not do it all the time?

In order to make new gestalts,

> You have to *risk* destroying old gestalts
> You have to *risk* facing the unknown
> You have to be able to bear your *impatience* when a new gestalt is not readily at hand
> You have to *risk* facing *anxiety* when your new gestalt is temporarily blocked and you feel the inadequacy of the old one
> You have to *risk* your *fear* of looking for a new integration when you are uncertain that you will be able to find it.

Sometimes the process of inhibiting growth is called "neurotic." But the same process that inhibits growth in one situation can lead to growth in another, so I want to beware of labels.

As the gestalt changes, so the meaning changes.

Let us begin talking about being yourself by talking about pleasure.

When you were a baby, and when I was a baby, a great deal of our world was based on feeling pleasure, finding pleasure, trying to keep pleasure from going away.

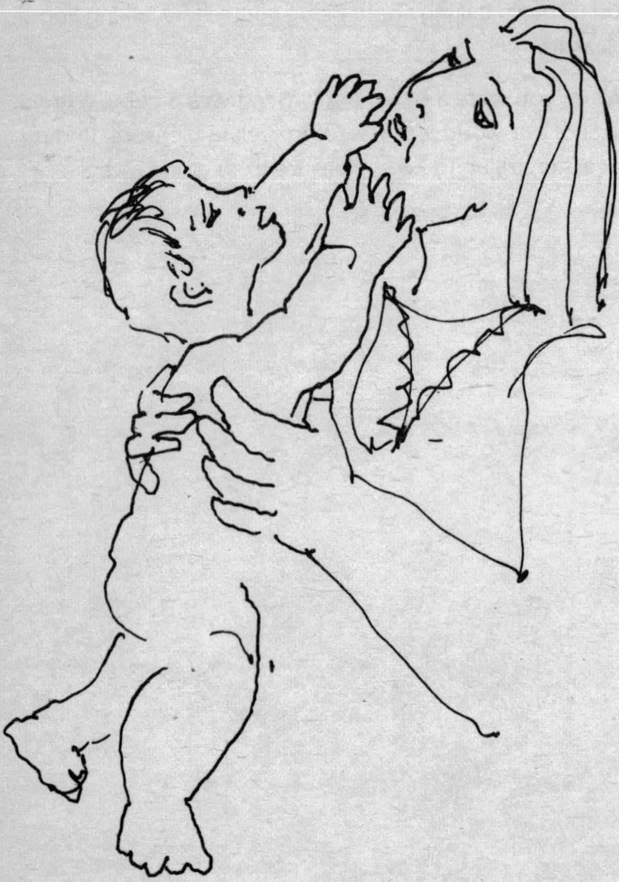

What kinds of pleasure?

>   feeling warm
>   being held
>   sucking on Mother's breast
>   looking around
>   playing with our fingers
>   feeling full
>   shitting
>   playing with our genitals
>   playing with our bodies
>   falling asleep

Now comes the sad part.

How many of these pleasures are still pleasures for you today?

I bet not very many.

What happened to your enjoyment of those early, powerful, simple pleasures?

Very likely you gave them up to please your parents. You started to think of them as childish, wrong, bad, or silly.

Or maybe you kept them, but they are now only to be enjoyed in secret, something to do and then feel guilty about.

Or to feel ashamed of.

Or to feel disgusted by.

In other words, you gave up pleasurable parts of yourself because you wanted to please your parents.

**How did this happen?**

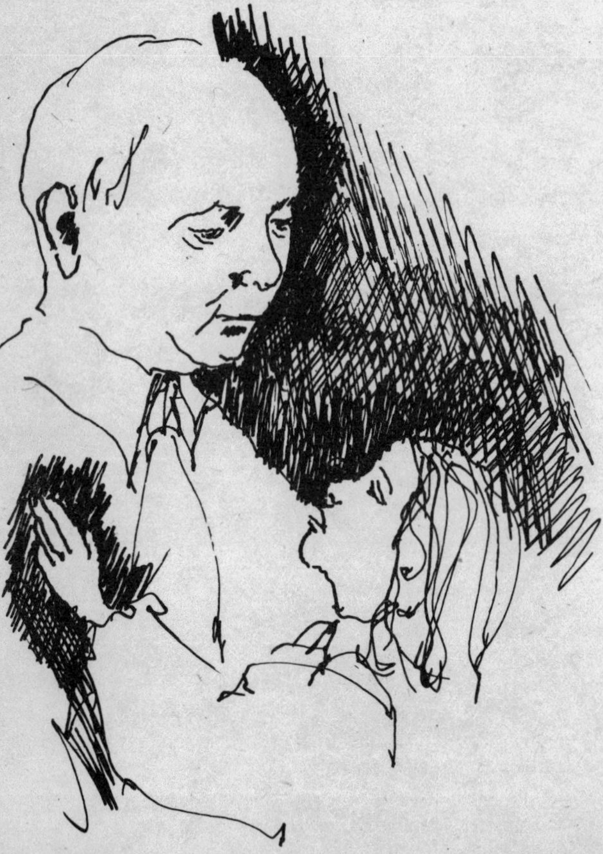

In order to get the pleasure of giving your parents pleasure—of showing what a good child you were—

> you gave up the early powerful, simple pleasures to get a new pleasure:
> Pleasing Mom and Pop

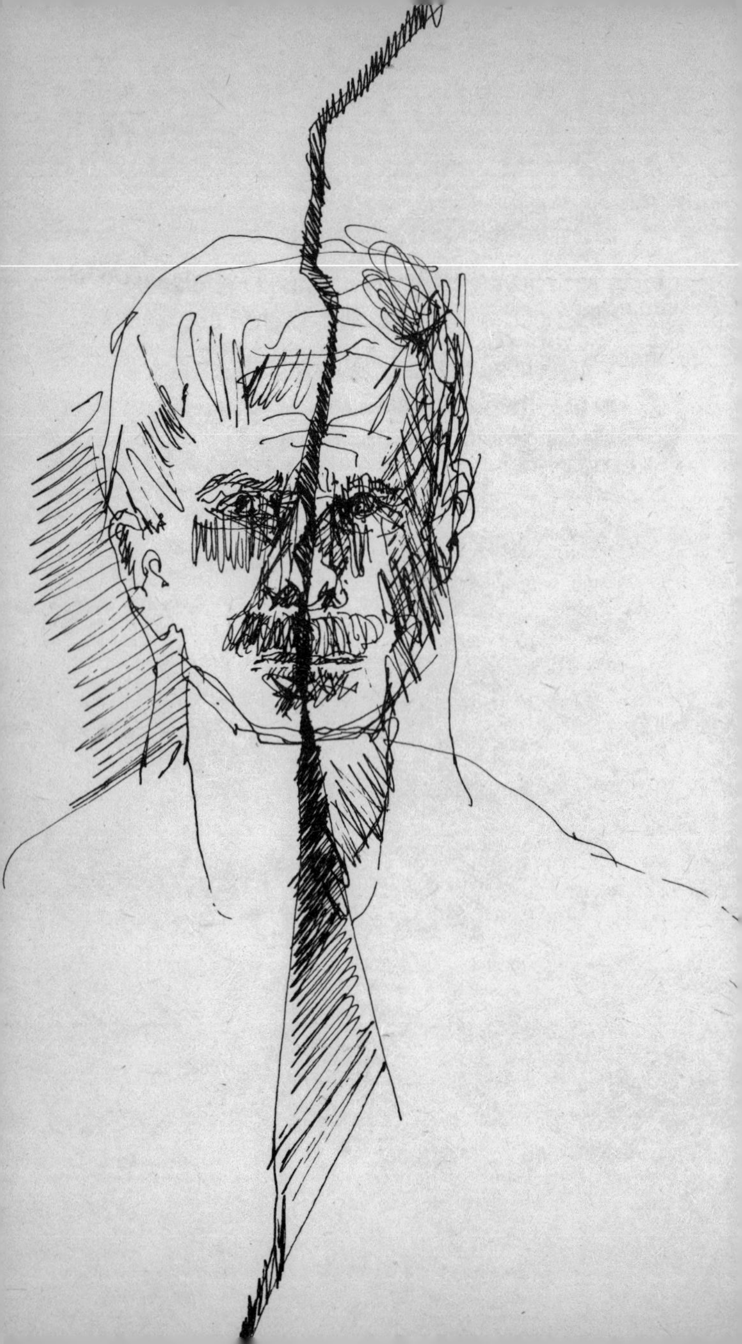

When faced with two different kinds of pleasure, you felt *conflict*.

Conflicts are painful:

> I want to be faithful to one person
> and I want to play the field

> I want many friends
> and I want to express my anger

> I want lots of money
> and I want to take it easy

> I want to win
> and I want to be nice to everybody

# right
# wrong

In order to resolve a conflict, sometimes I come to a premature solution.

I decide that one side is *right* and the other is *wrong*.

I decide to forget the wrong side.

But because the solution is premature, fake, unfinished, I have to spend a lot of energy making sure I don't think about the *wrong* side.

When faced with conflicts, the *good* person decides:

> to be faithful
> to keep his friends
> to work hard
> to control his competitiveness

In other words, to try to please his parents.

But, the early, young, simple, powerful pleasures are still there, still alive.

Only now he cannot allow himself to think about them.

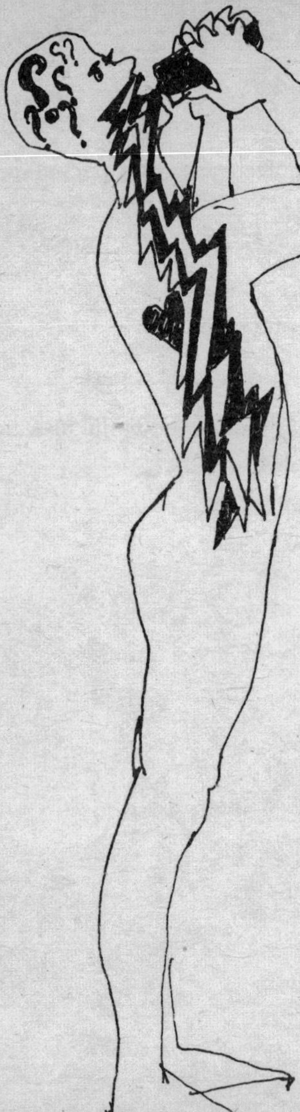

In order to

>   avoid conflict
>   reach a solution
>   please parents
>   feel mature
>   feel superior
>   feel good

most people turn away from many important pleasures that they are not finished with. They interrupt their pleasure or they may even block themselves from knowing what pleases them.

And they learn to do this by first *swallowing* someone else's *idea* of what is

>   good and bad
>   right and wrong
>   innocent and guilty
>   virtuous and sinful

And now that they have swallowed what is indigestible, they continue to have

> stomach aches
> diarrhea
> headaches
> "loggy" feelings
> disgust
> nausea

If you have swallowed someone else's ideas about pleasure,

>   then you have interrupted your own
>
>   > growth
>   > development
>   > excitement
>   > pleasure

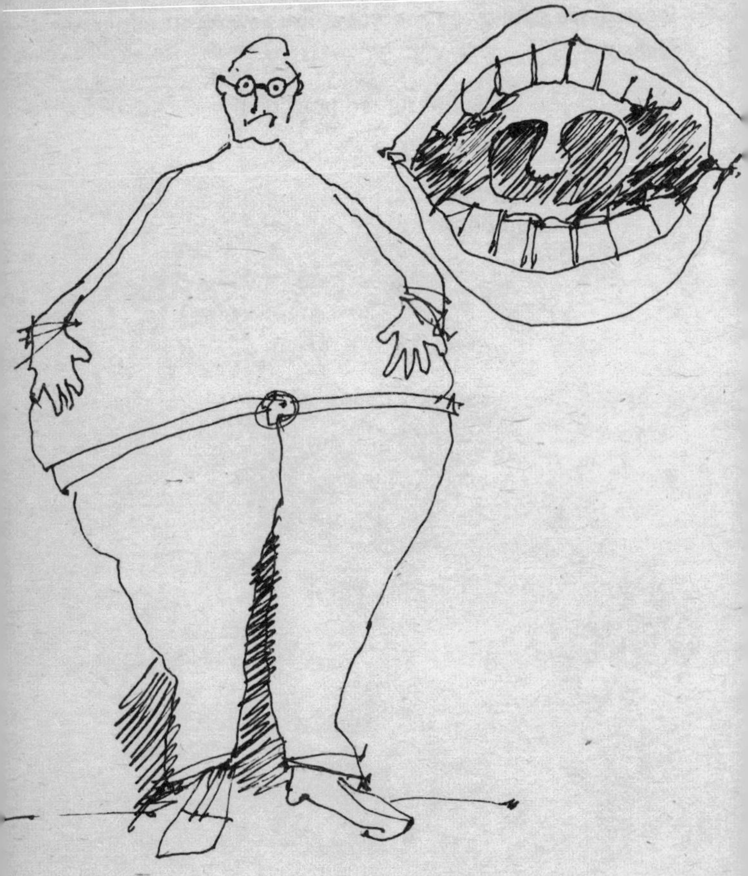

If you are to get back your own sense of pleasure, of feeling good, of feeling whole,

If you are to find out how you feel today, not what you have been taught to believe

If you are to continue to grow,

> then you will have to go back and give up what you have *swallowed* and learn how much you have *digested*.

Most people are *not aware* what and how much they have swallowed. Too often they take what they have swallowed as what is natural.

One important part of my work is to help you find out what you have swallowed without thinking about it, that still is a pain in the stomach.

One good way of becoming aware of how much you have swallowed is to pay attention to how often you use the following words:

>    ought to
>    should
>    have to
>    must
>    always
>    never

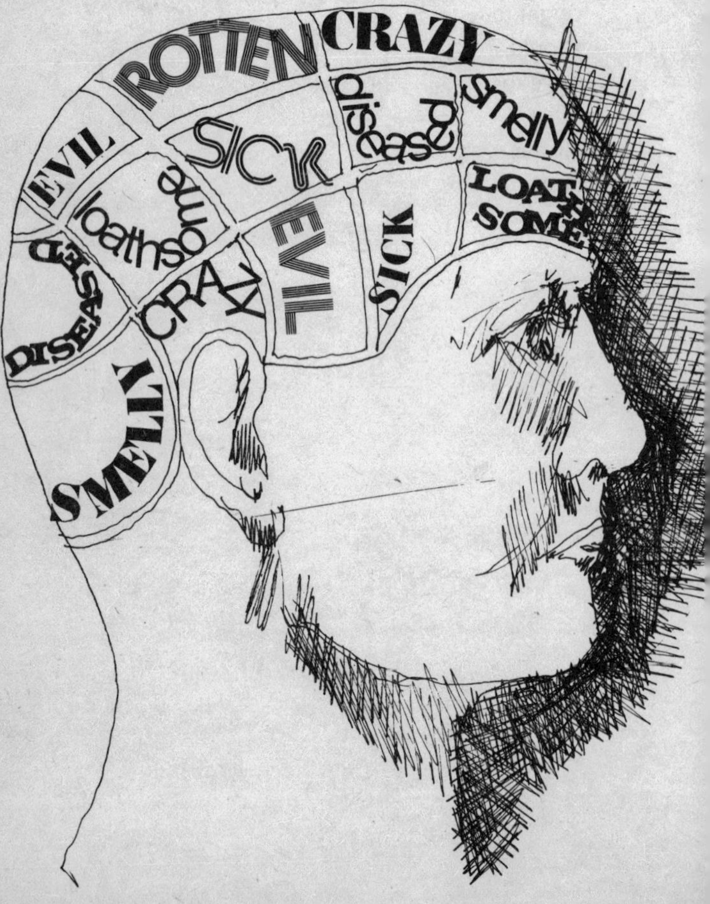
CRAZY
ROTTEN
EVIL
SICK
diseased
smelly
loathsome
LOATHSOME
SICK
EVIL
CRAZY
DISEASED
SMELLY

Other words that suggest swallowed ideas are:

- crazy
- dirty
- sick
- rotten
- evil
- smelly
- loathsome
- diseased

Sex is one place where we are given lots to swallow. So let's use sex as an example of how we can become more ourselves, instead of who we are supposed to be.

Let's make up a whole series of sentences about what people are supposed to be like sexually.

> I ought to be pure and clean.
> You should never masturbate.
> I have to remain a virgin until I marry.
> You should always shower before and after having sex.
> Sex is so smelly and sweaty.
> Only a sick mind would think of anal or oral sex.
> What a diseased book this is. Full of disgusting sexual ideas!

Do you have a lot of feelings about these "swallowed" words and their connection with sex?

Let's try some more.

Let me offer some swallowed ideas as if they were holy, as if they came from God, as if they were divine principles to be followed without thinking.

> Sex outside of marriage is wrong.
> Masturbation is bad.
> Homosexuality is evil.
> Having sex with the man on bottom and the
>     woman on top is perverse.
> etc.
>     etc.
>         etc.

Where did all of these ideas come from?

People who are easily embarrassed
> who easily feel guilt
>> shame
>> disgust

People who are righteous

People who suffer a great deal

People who try to be perfect

People who are looking for ideal situations

People who are too good, too nice

People with these feelings have probably swallowed more than they can digest.

If you feel like this,

> probably you have swallowed too many ideas that are still not a real part of you.

> It might be useful for you to reconsider what you have swallowed. This is not always easy. The reason you must face up to what you have swallowed is that you have some kind of indigestion.

Yet you may prefer to inhibit your growth rather than face the unpleasant feelings that may go along with reexamining what was once so easily swallowed in order to gain parents' approval.

Sometimes children try so hard to please their parents that they anticipate how their parents will feel about something; and they don't bother to check it out.

Like: "I just know my parents will disapprove of masturbation, so I will feel guilty about it."

Some parents won't, but the child has already gotten a premature answer and is stuck with it.

One of the good things about guilt is that you can do something you want—play hooky, peek at your parents in the bathroom, smoke cigarettes—and then pay for your pleasure by saying, "Oh, I'm a bad person. I punish myself by being guilty. You see, I really am a better person than the things I do."

DISGUST
SHAME
FEAR
ANXIETY
ANGER

Now here comes the hard part.

If you want to challenge the ideas you may have swallowed,

> then you will have to be willing to deal with your
>
> > disgust
> > shame
> > fear
> > anxiety
> > anger

To get a new gestalt,

> you will have to be willing to take all of the sentences that deal with sex in the past few pages and try to reverse their meaning.

For example:

> I don't want to be pure and clean.
> I like to masturbate.
> I don't care to shower before or after sex.
> I think premarital sex is just fine.
> etc.
>     etc.
>         etc.

If you feel uncomfortable, just stop.

The point is clear enough.

Many ideas about sex, which have been swallowed, remain inside us and stop our excitement and growth.

In the same way, many swallowed ideas about

> justice
> law and order
> government
> health and sickness
> man's spirit
> man's fate
> diet
> cleanliness

keeping us from growing. But if you want to get beyond being stuck with swallowed ideas or feelings, you may have to face other unpleasant feelings. Part of the price you pay for—preferring swallowed ideas to finding your own way—is that you can avoid conflicts. You have to decide for yourself whether you prefer to be stuck *or* if you want to risk unpleasant feelings and grow up.

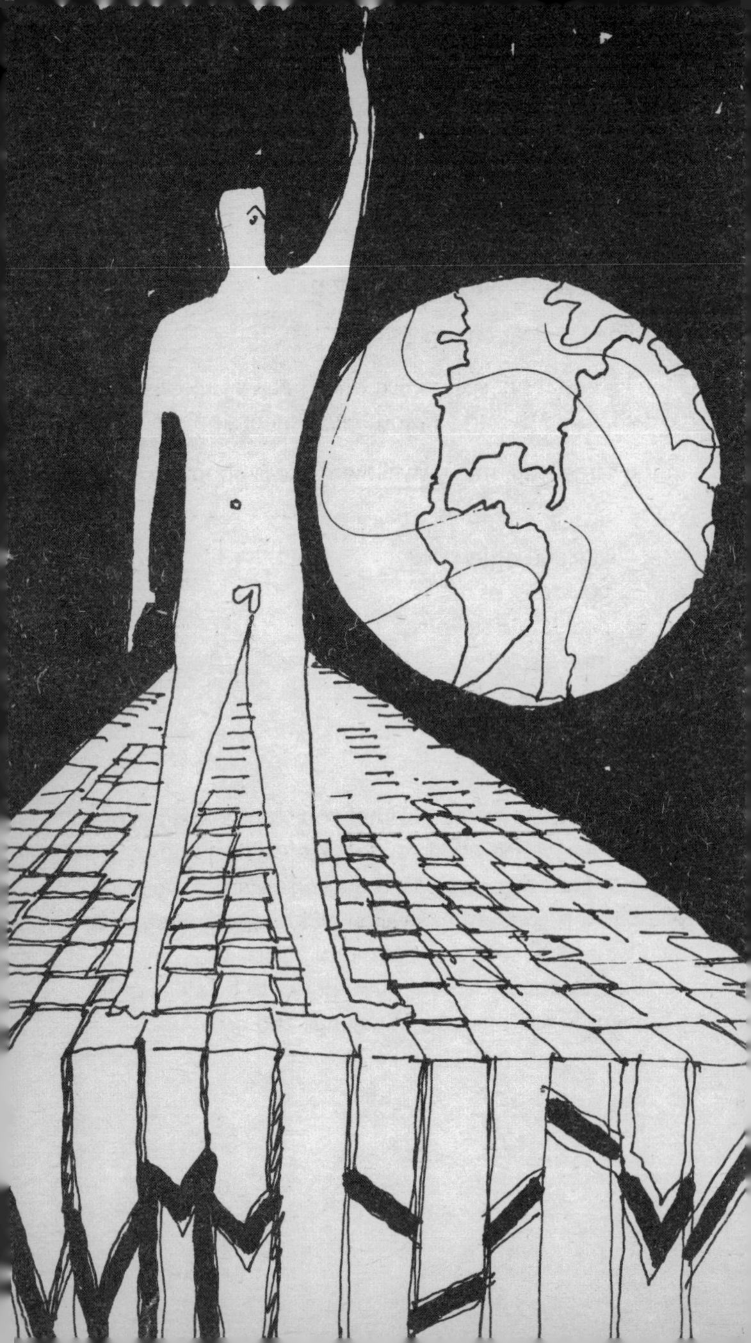

Do you want to risk finding out for yourself

>    what you like and don't like
>    what feels good and doesn't
>    how you want to live your life for yourself

Can you look at your life and decide

>    who you are
>    how you feel
>    what you like
>    what you want
>    whether you want to try to get it
>    what you can settle for

And so, for each of you, your answers are your own, just for you, and you make them for only one person.

Not to fit everyone into them, not for the whole universe. Just for you.

# 4

## What I Can Learn About Myself When I Get Angry and Blame Others

In the last chapter we discussed ideas and feelings that you swallowed from others and which gave you indigestion.

Here we are going to look at ideas, thoughts, and feelings that you do not want to accept, which you try to disown, which you try to claim are not yours.

*Why* do you do this?

Well, I can't be sure. Gestalt therapists are never sure about questions that answer *why*. But I can speculate.

In order to stay on good terms with myself, I like to think of myself as a good, just, gentle, sweet, loving, trusting person.

(But since I live with myself, I must have a pretty good idea that this is not always the case.)

LOVING
KIND
GENTLE
TRUSTING
SWEET
GOOD
JUST
REASONABLE

Alone at night

I sometimes admit

maybe I am not so

> loving
> kind
> gentle
> trusting
> sweet
> good
> just
> reasonable

But only then.

And sometimes in order to admit all these things, I have to be asleep.

And these feelings and ideas have to come out disguised in dreams.

I am afraid of accepting these parts of myself, so I try to lose my own connections with me here and instead

I GET ANGRY AND BLAME OTHERS FOR THEM

That makes me feel better, for then I don't have to admit that these "bad" feelings or thoughts are a part of me.

*I* am good. It is *you* who are bad and have these dirty thoughts and dirty feelings.

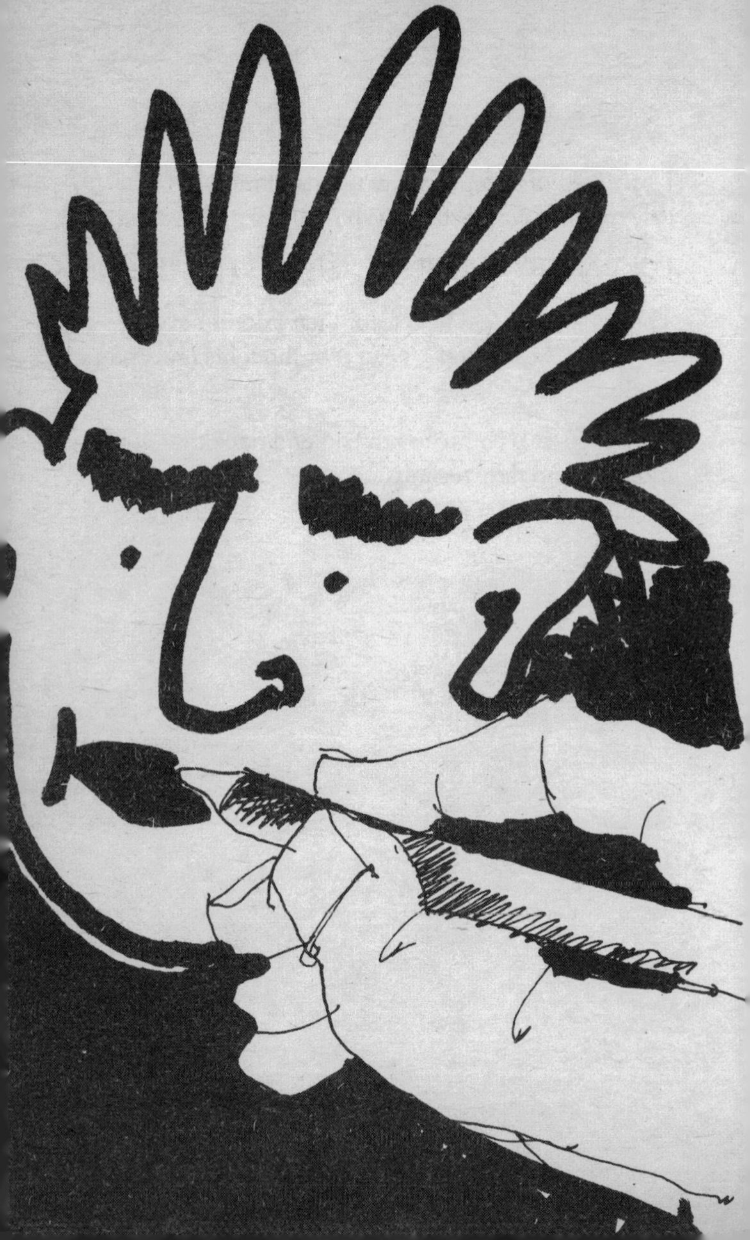

This is a hard chapter. Keep that in mind as you work through these experiments. Remember we are trying to discover what you gain from getting angry and blaming others.

Here's a hard experiment:

> Think of the person you hate most in the world.
>
> Conjure him up. Feel how disgusted you are in his presence, how you want to get away from him.
>
> Tell yourself all the things that you find so hateful about him
>
> Get all of your feelings about him out.

Now let's try something else.

Think of the various ethnic groups in America.

Think about them as

> niggers
> kikes
> wops, dagos
> krauts

And now I want you to apply to them all the traits you can't stand. Pretend that you are a bigot. Most of us are to a lesser or greater degree, though we don't like to think of ourselves that way.

Don't hold anything back.

Let it all out.

Now rest for a minute.

How do you feel?

Are you all puffed up with anger, all full of blame, black with rage, feeling rotten about all these ugly feelings?

Then that is what I have asked you to contact in yourself, and you can also feel proud of following my instructions. They are not easy for many people.

See if you can get something out of risking bringing up all these feelings.

Now let's go back to the first experiment: the person you hated.

Do you hate him because:

> he is two-faced, dishonest
> he is a dirty crook, a cheat
> he is mean and cruel;
> he is always angry, grousing, disloyal, gossips about you;
> he is cheap, penny pinching, drinks too much,
> he is sexy, too sexed up, lazy;
> he tries too hard to please,
> he is too smart for his own good,
> he is too dumb to be endured,
> he spends money like it was going out of fashion,
> he never loosens up and gets drunk,
> he smiles too much;
> he smiles too little.

Now go back to your bigoted feelings.

How many of them are the same or similar to your feelings about the person you hate so much?

Quite a lot of them?

Think about it.

(Remember in the first chapter, I asked you to try not to judge, but to accept what happened when you did an experiment dealing with the here and now?)

Now let's try another experiment.

Think about the person who knows more about you than anyone else. Maybe there is no such person, so invent him.

This person knows all the embarrassing, rotten things that you have done, but that no one else knows. He has the goods on you, the real dope, all your secrets.

Now, what are the worst things that he can say about you?

Think about it.

Isn't that interesting?

Now take all the things you hate about yourself and compare them with what you hate about the person you loathe and the minorities you detest.

How much of the list overlaps?

Can it be that all three lists are virtually the same person?

>YOU

What you are protecting yourself from, in all three instances, is facing the parts of yourself that you do not like, accept, own.

What you are doing is hating someone for being the person you are afraid to be.

Can you recognize that you may hate someone else for what you hate in yourself; that you hate someone else for what you fear or lack in yourself?

You are losing parts of your own feelings, parts of your own ideas.

You feel more comfortable seeing them as belonging to someone else, something else.

Which parts of the list have *nothing at all* to do with you?

Are you sure?

Positive?

O.K., then put those traits aside.

Now how much sense does all this make?

How much of yourself can you recognize?

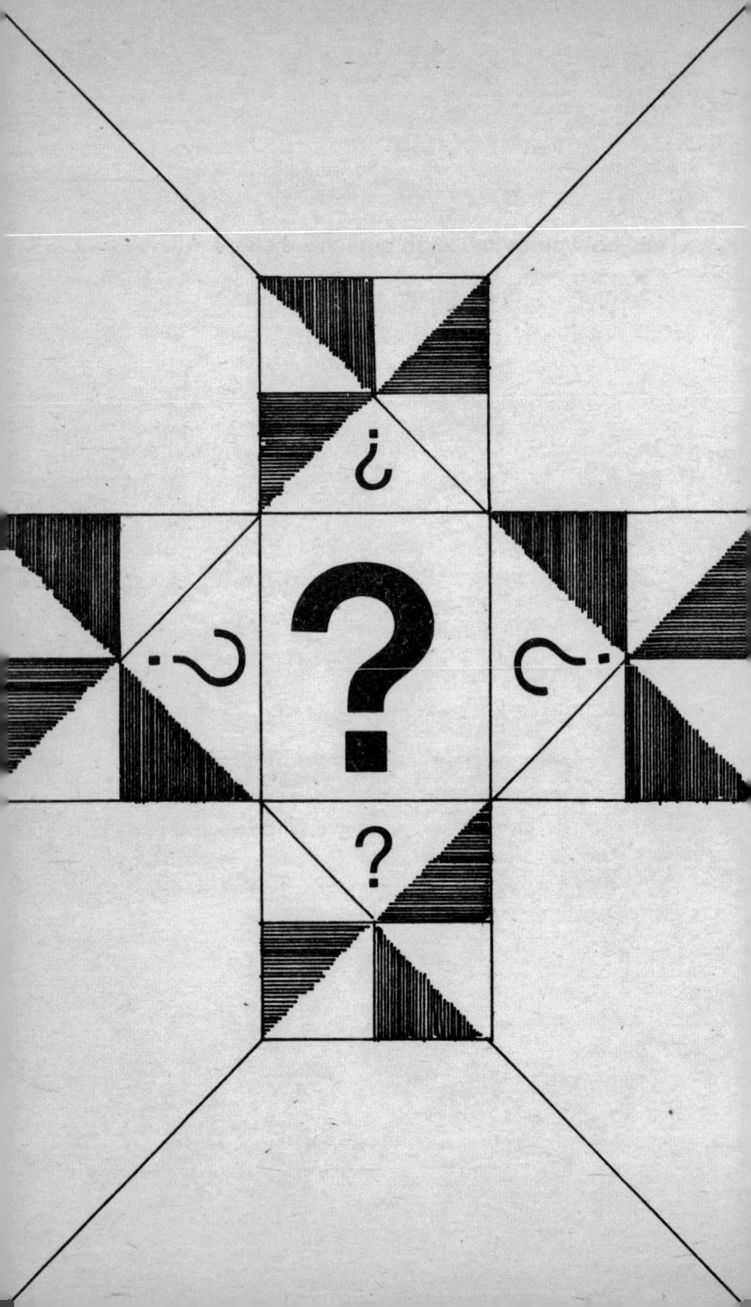

I warned you at the beginning of this chapter that it was difficult. Is it too hard?

You see this chapter deals with *projections,* a very serious way of interfering with how you live your life, how you ruin relationships.

Maybe you can't make too much sense of these experiments the first time you try them. That is O.K.

Try them again another time.

Let them rattle around in your head for a while, but try not to worry about them.

Remember, you are getting just a taste, a slice of gestalt therapy, and not the whole loaf.

If you become more open to yourself through these experiments, that doesn't mean that you have to finish it all up. You can do as much or as little as you like. You are in charge.

Of yourself. Of how you use this book.

When you think and mull over the experiments in this chapter, you may become aware that some of the important difficulties in your life arise not because others are bastards, but because you are angry and blame them for what upsets you about yourself.

If you can grasp this, then even if you still find some of the experiment confusing or difficult, you may also be thrilled to recognize that sometimes it is not that the world is so nasty or rotten, but that you have a lot to do with creating your own world.

That can be threatening, but it can also be liberating.

Let's go on.

Now try something else.

Take each of the traits you objected to in others and now try to make a case for it, try to justify it, try to see when it is useful, when it is a good thing to do.

Can you remember my list? You hate him because:

> he is dishonest
> he is a crook
> he is mean and cruel
> he is too smart for his own good
> etc.
>     etc.
>         etc.

Let me help you

Sometimes it's good to be untruthful because:

> You can protect yourself against people who are trying to cheat you.
>
> You can make people like you. Everybody needs friends, even you.
>
> You can prevent hurting people's feelings. Honesty can sometimes cause a lot of suffering.
>
> Nobody needs to know *that* much of my business. I like my privacy.
>
> etc.

Remember, I suggested, I hate him because he's too smart for his own good.

Now what about making a case for:

It's good to be very smart because:

you can protect yourself
you need to understand what others have in mind
you have a better chance of winning
you can get your own way more often
you can think through problems
you make fewer mistakes
you can figure out how to make others like you
etc.
    etc.
        etc.

Now try something else.

Try to see if you can say I am proud:

And then put down all the traits that were just so unacceptable to you.

Go ahead. Just try to say it, even if you think you can't mean it, just go ahead and say:

> I'm proud of being dishonest
> I'm proud of my ability to cheat
> I'm proud to be mean and cruel

Just continue with the list.

Your list. Not the list I have tried to provide.

Try it with both lists.

And then see what happens.

Here is another way of getting at the same thing. You may find this experiment easier and more fun than the last.

I want you to pick out the one thing in your home that you like best, that pleases you the most. Take a careful look around, don't leave anything out. Take as much time as you need.

Now list what you like best about it. What are the qualities you like?

Now I want you to pick out the thing you hate, that you want to throw out.

Do the same thing—think of all the things you don't like about what you picked.

Can you consider that the qualities you listed in the thing you liked are what you like about yourself?

Does this description fit what you care for in yourself?

How about the second list—the thing you don't like. Is this a list of things you don't like about yourself?

We often lose what we feel about ourselves—both good and bad—and then find it in others, without recognizing how it got there.

We put it there, or we looked until we found it there.

Go back to the objects you picked to like and hate. Notice that they have other qualities than those you chose.

Would it be interesting to know that a good deal of what the world is like comes from what you are looking for, the answers you already have, but keep finding out there? Just consider that your picture of the world out there is a self-portrait.

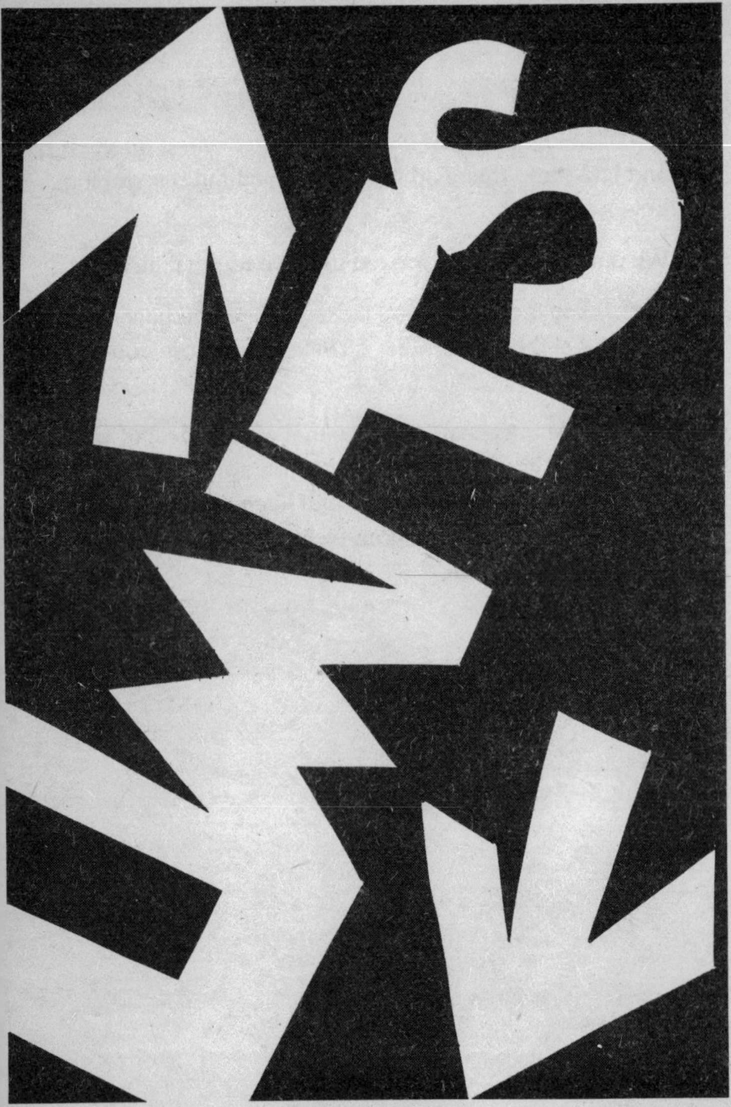

Whatever you *feel* when you try to recover lost feelings and ideas is also an important part of what is happening to you right now.

What you are *feeling* at the same time as we look at *what you have lost* then helps us to take the next step in recovering the part of you that is lost.

In other words, we are working to change your gestalt of yourself. We are trying to enlarge your understanding of you, even if it means coming to a limit.

We all have our limits, and we all have to come to terms with them.

One important part of being with myself (who else can I be with?) is I have

    to accept my limits

        (what I can't change
            or what
      I don't want to change)

        or

    to accept my possibility of change

        even if it means
            SOME
                hard work
                pain
                tension
                fright
                etc.

Therapy involves all these things.

And although I can help you so that you may not have to deal with more

> hard work
> pain
> tension
> fright

than you are willing to accept (that is, I respect your limits), if you want therapy, you probably will want to risk facing these unpleasant feelings and proceed.

Some parts of therapy may be fun and thrilling, and some may be sad and painful.

Now this is also hard to accept, but much of what therapy is depends on *you*, what *you* want it to be.

You create in large measure what your therapy will be like.

One limit of your therapy may come from my own limits.

But no therapist likes to talk about that.

Not even me.

But if you can try to accept me,

as I try to accept you,

    then I can also try to recover these parts of myself.

# 5

## How I Can Avoid Turning Against Myself and My Body

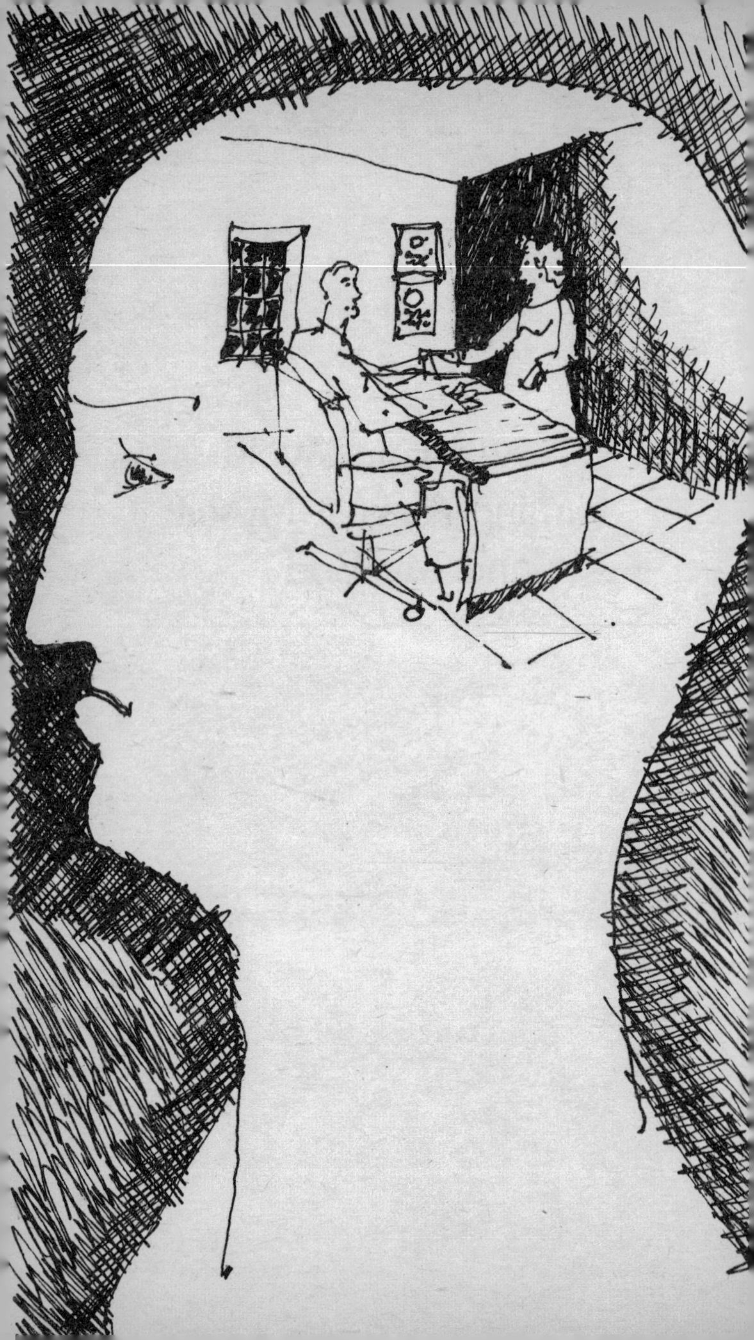

Many people want to forget about their bodies.

They want to live in their heads.
They want to live above the neck.
They forget about their hands. They are clumsy; they just don't work.
They want to forget about their torsos and limbs. They are weak.
They want to forget about their genitals. They make too much trouble.
They want to forget about their asses. They don't want to know that they shit.
They want to forget about their lungs. They don't want to be bothered about breathing.

But we *are* our bodies.

We are our bodies, but we are not just our bodies.
We cannot exist without our bodies.

The reason *why* we try to get away from our bodies is something I can't prove. I can only present a point of view and some evidence.

You are free to chew it over and see if it tastes good or spit it out if it doesn't.

It's *your* life. It's *your* body.

You have to consult *you*, see how *you* feel, what *you* think, what *you* believe, what makes sense to *you*.

In order to live, I have to deal with the world, my environment.

But I am also a part of that environment.

If I breathe, I take air in and let air out. I am part of my environment.

If I love, I take someone, some thing, and I take it into me, and I do something to it. I add love, and, in this way, I change my environment.

In other words, *I* am always active, a part of my environment, making my environment, changing my environment, being restricted by my environment.

The bulk of my involvement with my self and my environment takes place either through

    my body

        or

    my personality

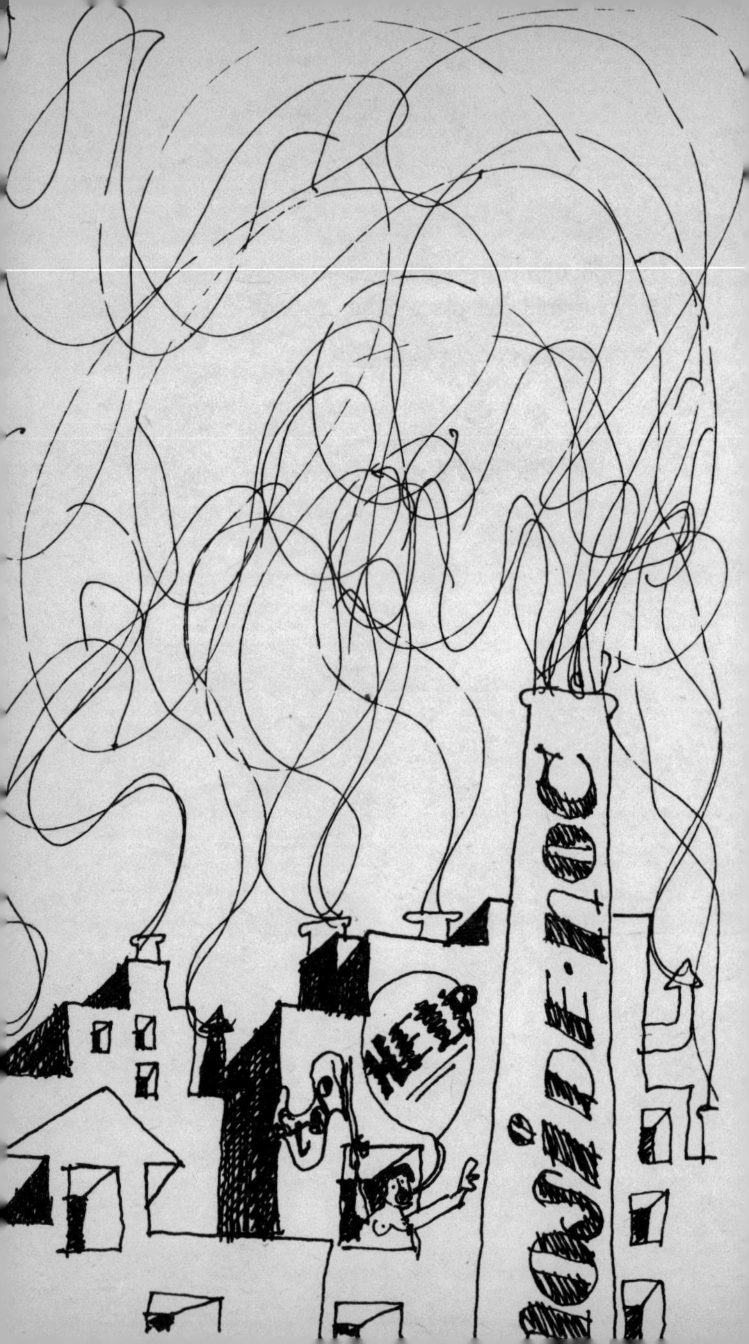

A great many of my contacts with my environment involve destruction.

A great many involve aggression.

> I breathe in good air and breathe out foul
> I take in food and turn out shit
> I want sex, and that seems often like rape
> I get angry, and that often seems like rage or murder
> I get bored, and that means I want to change what is happening
> I get disgusted, and that means I want to get rid of what is going on, to get away from it

I have learned to be

> ashamed
> disgusted
> embarrassed
> guilty

about

> my bad breath
> my body smell
> my shit
> the dirt under my nails
> the hair under my arms
> my long toenails
> my nose droppings

and certainly about

> my genitals
> my genital smells

Here's a piece of advice.

Straight, flat out ADVICE.

Whenever you feel

> ashamed
> embarrassed
> guilty

try not to let this stop whatever else is happening for you at that moment.

If you can accept your shame, embarrassment, guilt *at that moment and keep going,* all kinds of exciting things will happen for you.

Examples:

If you are meeting someone for the first time,

> try not to let your *embarrassment* keep you from getting to know that person

If you are visiting a new city or country,

> try not to let your *shame* or timidity at not speaking the language or your embarrassment at not knowing where places are become the means to not risking getting to know that city or country better

If you are reading a new book or seeing a new play,

> try not to let your *disgust* become the means to not finding out what is threatening you at that moment

If you are having sex,

> try not to let your shame, embarrassment, or *guilt* keep you from getting to know what other feelings are there, like pleasure, love, excitement, joy.

So, please be *aware* of your shame, disgust, guilt, embarrassment,

BUT DON'T stop there.

Accept your shame, etc., and go on and see *what else* is present at that moment.

My guess is that in addition to your shame, etc., you will find lots of other feelings to make it important for you not to be stopped by the negative feelings.

We first learned to be ashamed, embarrassed, or guilty when we were children. Babies are born without these feelings.

These strong negative feelings are a very powerful means of teaching children what is expected of them.

>   Don't wet your pants—only babies do that.

>   Don't cry—only girls, sissies, or babies do that.

>   Don't spill your food—only babies do that.

These powerful negative feelings become too powerful. You lose control over them and invoke them routinely, where they don't belong.

In this chapter I want to provide a chance for you to see how your powerful negative feelings lead you to turn against yourself and your body.

But there are more physical ways you can stop yourself from getting more out of your environment.

You can be using your body to stop yourself from doing to the environment what you have been taught is bad or frightening or dangerous or whatever.

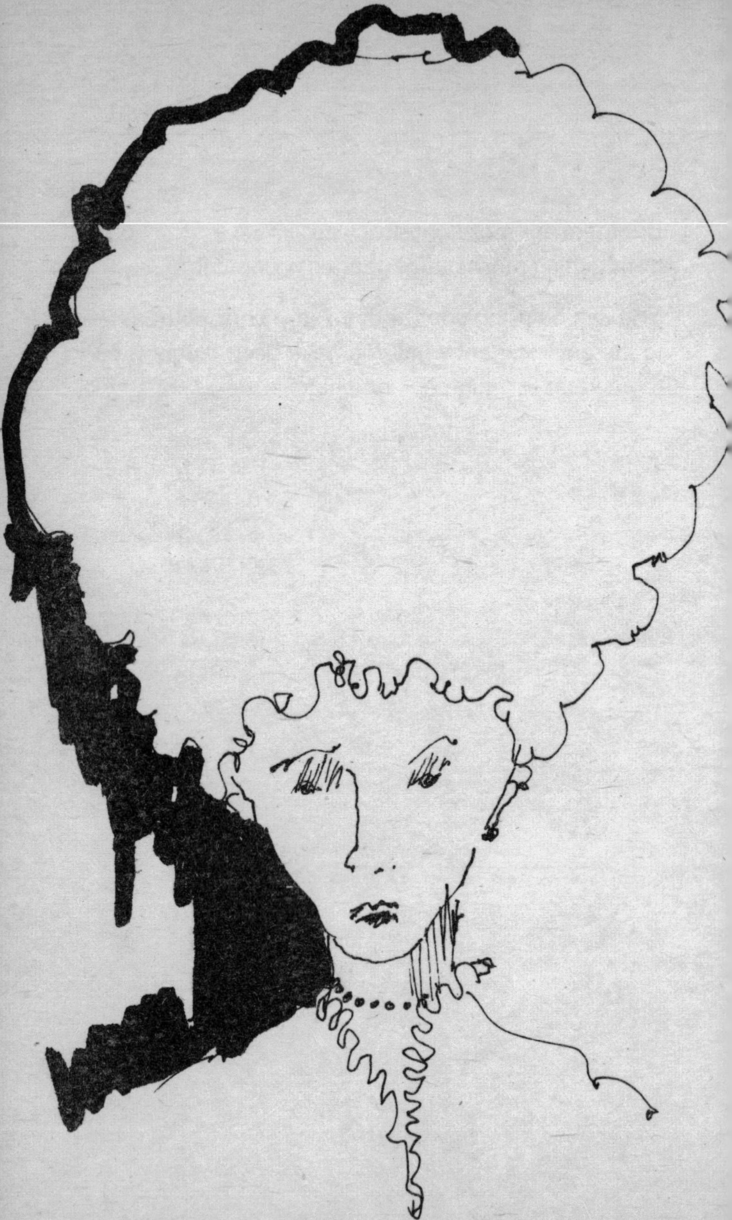

A few examples:

A young woman comes to see me, very proud of a painting she has just finished. She says that her head feels swollen. It is too big for her body.

Then she connects. She has a swollen head. She is not permitted to feel pride. She is ashamed to be proud of her painting, because she had been taught that she should not boast or brag.

Her head now feels like it has its own size.

She has done to herself, to her head, what she could not do in her environment: to feel big, larger than life, to be proud of herself without shame or embarrassment or guilt. Instead she chose to punish herself for her pride.

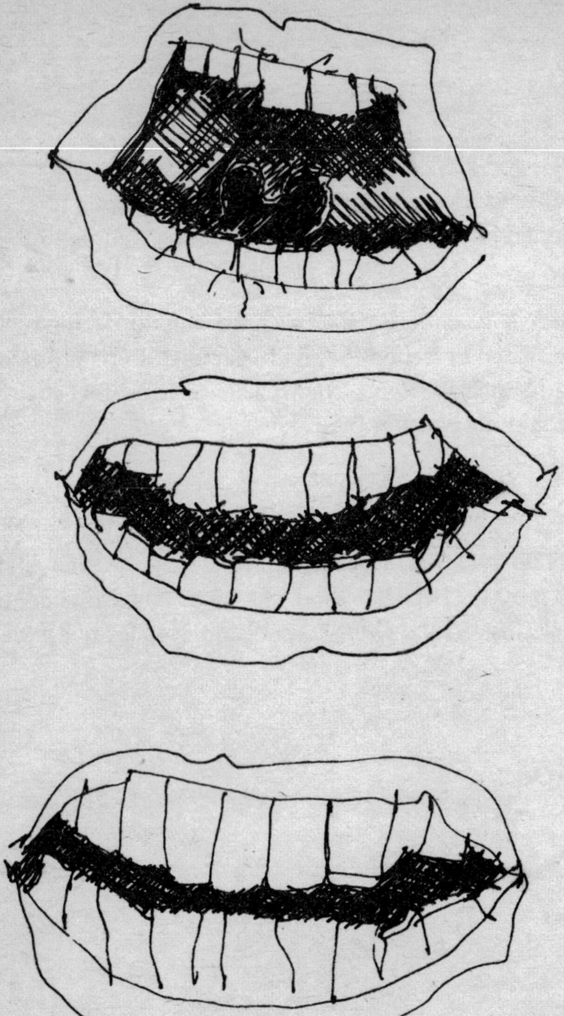

A young man tells me that he grinds his teeth in his sleep. He asks me what it means, what to do about it.

I ask him to grind his teeth right now. To exaggerate the grinding. To tell me what he feels.

He does so. He grimaces. His jaws work furiously.

RAGE.

He is angry because he has swallowed many of his parents' ideas as to what a good young boy does. He does not permit himself to feel his anger (even though he manages not to please them) and instead he grinds his teeth.

(As he talks, he spits out his words.)

(I want to help him feel his rage and not be afraid that if he does so, he will be a murderous bully. He fears that what he felt as a child is what he will feel now if he permits himself to be aware of his rage.)

A middle-aged housewife suffers from migraine headaches. She has had them since she was eight years old. Her father and mother had them. They are inherited, so she says. She consults the best physicians in the world on migraines and she takes a whole medicine chest of drugs.

Still the migraines come.

I am not sympathetic.

I think a lot of pain or suffering are bribes by which the person suffering often tries to blackmail his friends or family or himself from facing his feelings.

Right now, as she is experiencing one, I taunt her with these attitudes toward her migraine. She gets angry, calls me a bastard, reaches out to punch me.

And lo, her headache is gone. She feels a little frightened, a lot angry, and quite alive.

(She learned long ago that she was naughty if she felt angry, so now she turns her anger against herself and experiences it as a migraine headache. She has turned her anger, her shame, and guilt over it, into a headache.)

A young woman is angry because her lover has gone off on a trip without her. But she loves him and cannot allow herself to be angry with him. So while he is gone, she develops a migraine headache and terrible stomach cramps. When he returns, her headache and her stomach cramps stop. But the next week, when he leaves on a business trip, they return. Yet she refuses to talk to him about her loneliness, jealousy, anger.

Now I want to turn your attention to your own body.

As I said at the beginning of this chapter, we are our bodies (although we are not just our bodies).

Too often, we take our bodies for granted. Forget they are there. Ignore them.

Stop all of that. Right now.

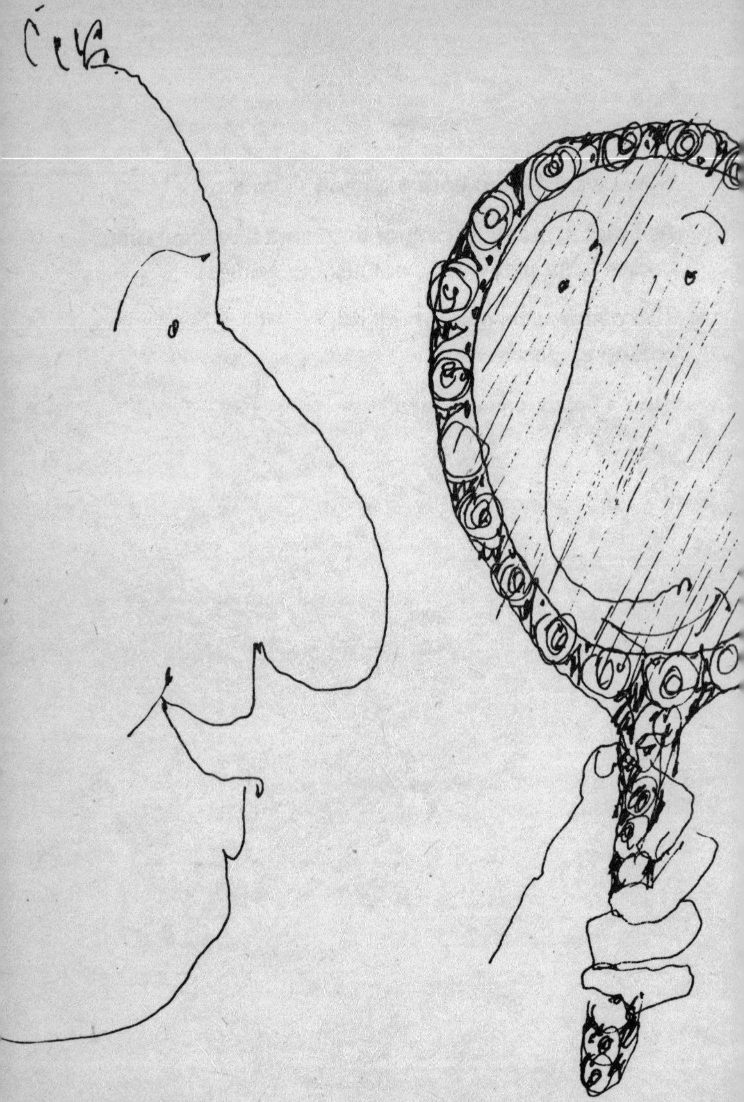

First, concentrate on your body.

Think about all of the things you don't like about your body. For instance,

>    your waist is too fat
>    your legs are too short, fat, thin, or long
>    your hair is too curly or straight
>    your nose is too long or short

Most people who don't like certain parts of their body really are talking not about their bodies but THEMSELVES.

Try something else.

Take all those parts of your body you hate.

Try to find ways of

> admiring them
> loving them
> being proud, even vain, about them

Now try something else.

Think and feel about all the parts of your body which

>   you like
>   are proud of
>   admire

Turn these into their opposite.

See if you can hate, condemn, distrust, those parts of your body.

Now see how you feel about yourself and your body.

I want you to try something different.

A simple experiment.

Simple only in terms of the directions. Let's see if it is simple *to do*.

Lie down on the floor and see that you have lots of space to move around.

Once on the floor, just lie there. Feel the floor under you. Let the floor hold you up.

Feel the floor under your ass, holding you up. Let go, and feel the floor holding your head. Let it rest on the floor; don't hold it yourself.

Feel your arms resting on the floor, just lying there. Feel the floor underneath them, holding your arms.

Now do this with all the parts of your body. Your chest, your legs, your feet.

And notice your breathing. First just see how you breathe. Pay attention to your breath.

And then, when you see how you are breathing, see if you can push all the air out of your lungs so that they will fill up. Continue this deep breathing.

So far,
all of this will take about two minutes.

But I am greedy. I am greedy for you to have as much as possible.

So please go along with me, no matter what you are feeling, unless it is extreme discomfort. If you feel very uncomfortable, then just stop.

Now if you are still with me, just continue lying on the floor quietly, paying attention to your breathing and feeling the floor under all of your whole body.

Do this for at least a minute.

Now comes the hard part.

As you lie on the floor, breathing deeply, I want you to pay attention to the slightest feelings in any part of your body.

No matter how small, infinitesimal, vague.

And as you feel those tiny, tiny, small feelings in your body, see if you can find what they want to do, whatever small, tiny movements you want to make.

And without any order, without logic, without any purpose or meaning, just pay attention to any small part of your body that wants to move. And go along with it. Let yourself move that part of your body.

And let those movements keep going. If you want them to get larger, let them. Let them go in any way they want. And then when you feel finished, STOP.

What happened?

It all depends on you. If you liked what happened, try the experiment again.

If you didn't like what happened, try the whole experiment again.

Now what happened?

If you were free enough to go along with whatever feelings you had in your body,

then you might have released a lot of the tension you usually carry around,

you might have let go of a lot of squeezing of your muscles, which you do without even thinking about it.

Yet you might have been unwilling to go this far. And this is O.K. For if you think about what you did with the experiment, you can recognize how you relate to your body. And this is important to know.

> Did you distrust it and just lie still?
>
> Did "nothing" happen? (Please describe this "nothing.")
>
> Did you drift off after a few seconds?
>
> Did you get bored?
>
> Did you get tense?
>
> Were you delighted to discover the small urges toward movement that you felt?

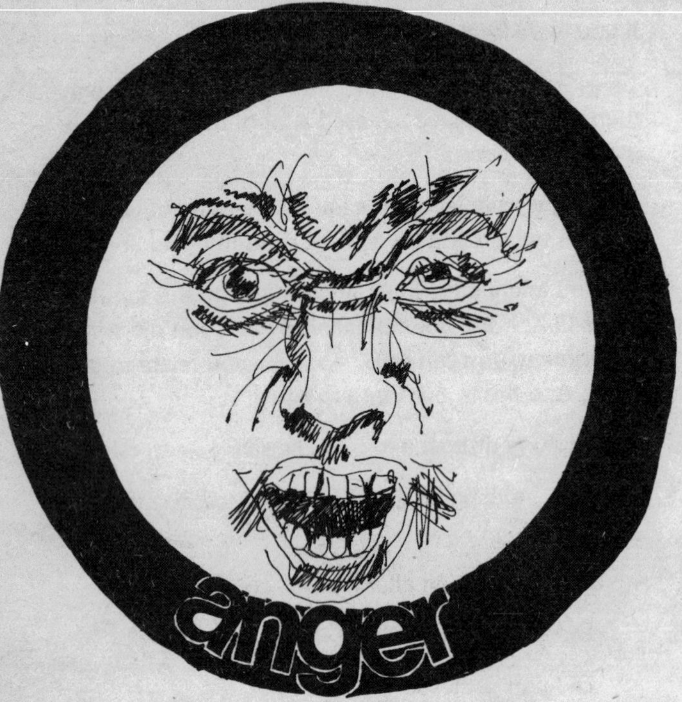
anger

Now please try one last experiment.

Think of someone or something that made you angry *today*.

By *angry*, I mean the whole range of annoyance, irritation, dislike, rage, fury.

Now

    get a pillow or a cloth

If you get a pillow, pretend it made you angry, and pound away at it. Be sure to breathe deeply.

If it is a cloth, pretend it made you angry, and while remembering to breathe, strangle it, throttle it, garrote it.

Really get into it. And if you want to scream or shout at the same time, go ahead!

Do it! Now!

If you can really let yourself get angry and express it toward the pillow or the cloth, then you are feeling lots of strength, power, excitement, refreshment.

But if you did not let yourself do it, then you are more aware than you were before of how important it is for you to keep strong control of your anger, even against a pillow, a cloth.

And you may also be aware of some of the price you pay for it.

Please *do not misunderstand* what I am doing here.

If you can express your anger toward the pillow, I am *not* suggesting you do the same thing to whatever it was that originally made you angry.

I am *not* saying go out and assault whatever or whoever made you angry.

But I do want you to get an idea of how much energy you use for controlling your anger.

I want you to see how much you may deaden yourself by keeping the lid on.

I want you to recognize how much tension builds up through so much control.

I want you to see how much excitement gets blocked.

And if we were working together, I would want to explore ways in which you could appropriately express your feelings without turning against your own body,

I would want to explore the possibilities open to you of relieving your tension,

of accepting all of you: your feelings, your thoughts, your body.

It is pretty important to do.

How do *you* feel about it?

What do you *think* about it?

Do you feel you have already started?

# 6

## What Happens Now?

EXCITED?
HOPEFUL?
TINGLING?

NERVOUS?
FRIGHTENED?
ANXIOUS?

BORED
DEADENED
IRRITATED
CONTEMPTUOUS

FATIGUED?
heavy?
EMPTY?
drifting?

CONFUSED?
STUPID?

What happened to you as you read this book?

Did you try the experiments?

Did you find yourself feeling

>excited
>hopeful
>tingling?
>
>nervous
>frightened
>anxious?
>
>bored
>deadened
>irritated
>contemptuous?
>
>empty
>drifting?
>
>fatigued
>heavy?
>
>confused
>stupid?

Each of these feelings offers an opportunity for your growth and development.

It is your decision whether to take the next step

    to get deeper into each feeling

    to let each feeling develop, grow, lead you further

    to get through it

    to get inside it, around it

    to get beyond it

    not to get stuck with it

It is your life, and you are in charge of it.

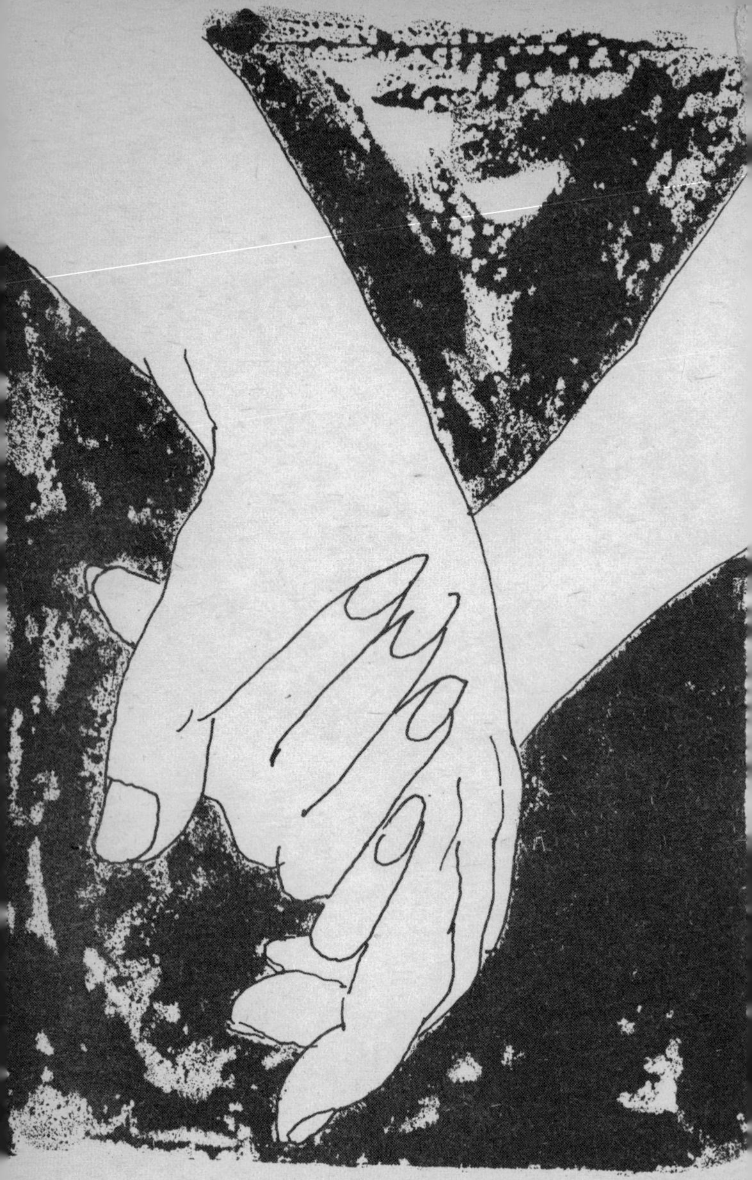

The experiences you had while reading this book are you.

You have the potential for other experiences.

How do you want to proceed to be fully you?

What happens now is up to you.

# Appendix
## A Brief Note About the History of Gestalt Therapy

Gestalt Therapy was first developed by Fritz Perls and his wife, Laura, about 1950. They were helped at that time primarily by Paul Goodman, Ralph Hefferline, and Paul Weisz:

Fritz and Laura Perls were psychoanalysts who were trained in prewar Germany. Later Fritz Perls was supervised by Wilhelm Reich. Earlier Laura Perls, as a graduate student, attended lectures by the gestalt psychologists: Wertheimer, Kofka, and Köhler.

Out of this background, and as a result of their experience, they became aware of the limitations of psychoanalytic theory, and they began to experiment with new ways of working with patients and thinking about therapy.

In the beginning, Fritz and Laura Perls thought about calling their new form of therapy "concentration therapy," "awareness therapy," or "existential therapy," but they settled on "gestalt therapy." During the 1960s Fritz Perls migrated to California and British Columbia while Laura stayed in New York. The major differences in gestalt therapy arise from their differences in theoretical orientation and ways of working with patients, as they continued to develop their own styles.